YOUR KNOWLEDGE HAS VALUE

- We will publish your bachelor's and
  master's thesis, essays and papers

- Your own eBook and book -
  sold worldwide in all relevant shops

- Earn money with each sale

Upload your text at www.GRIN.com
and publish for free

How Classism Can be Detected in the Sector of Mobility. A Suggested Method

Florian Wondratschek

Bibliographic information published by the German National Library:

The German National Library lists this publication in the National Bibliography; detailed bibliographic data are available on the Internet at http://dnb.dnb.de.

ISBN: 9783346682055
This book is also available as an ebook.

Nürtingen-Geislingen University of Applied Science

Sustainable Mobilities

Master of Science

Paper

Applied philosophy of science

winter semester 2021/2022

How does classism can be detected in the sector of mobility? A suggested method

Submission date: 18 December 2021

Florian Wondratschek

Content

Gender-appropriate language is used in the study

1. Introduction

The big SUV at the front door, the own car, the private jet at the airport or the yacht at the next harbour? Even at a time when the climate and environmental movements are on the upswing with Fridays for future and with the Greens entering the German government, such owns seem to be a lifetime dream for some people in our population. For many decades, it was true that if you drive up in a big and expensive car, you have made it in material terms - even if the vehicle is only leased or financed on credit, which is not apparent (cf. Knierim 2016, 52). Knierim (2016) is of the opinion that "especially in the pietistically influenced southwest of Germany" the Daimler or another limousine is considered a driving proof of wealth and righteousness (cf. ib.). It seems that for some, psychologically, one's own vehicle functions especially as a status symbol. In Maslow's hierarchy of needs, status plays a major role, but its definition via a vehicle takes place as an object. The status is externalised, according to Marxian thinking a kind of alienation. Is this appropriation process relevant to society? There are critical tendencies from sociology because it comes to a desired equality of people. For example, ways of thinking in which status symbols are acquired only on the basis of prestige ensure a separation from people who cannot afford them. This socially divisive materialised superiority thinking was given the social science term "classism" in the United States around 1900, which denotes prejudice or discrimination based on social status (cf. Kemper/ Weinbach 2016, 11). The term classism is formed similarly to sexism and racism to understand discrimination not only on the basis of ethnic or gender, but also on the basis of membership in a presumed class. Often this classism is related to an overuse of space that is harmful to the climate, which in the mobility sector is often demonstrated by the size of the car, the parking space or other luxury features. Classists individuals have an urge for status symbols, which is reflected not only in the mobility sector, but also in real estate issues, cultural issues and especially in the choice of partners (cf. Jensen 2012; Leondar-Wright 2005; Wondratschek 2020). Discrimination makes a significant contribution to an appropriated mindset of belittling and devaluing people (cf. Kemper/ Weinbach 2016, 51). It can be understood as a fundamental legal task to combat classist basic attitudes. First of all, however, the presumed need would be to survey how much classism can be found. In practice, revealing such attitudes is a major challenge for the social sciences, as their research can reach ethical limits when it "endangers the psychological well-being of participants" (cf. Scheytt 2020, 3). When considering the cause of classism, chronic inferiority complexes may play a role, which could trigger the subjects (cf. Häberlin 1947, 7). On the other hand, in order to avoid accentuation, socially desirable answers could be given, if it is made transparent that the research is conducted on personal tendencies towards classism (cf. Bogner/ Landrock 2015, 3).

Within this framework, this study aims to show how to detect classism, also in the sector of mobility, and thus create a basis for scientific research.

2. Classism as a field of tension

2.1.   How classism is a problem

The possession of status symbols was described by Max Weber (1980) as commonsality. It expresses hanging out the privileged position and appropriating a characteristic consciously or unconsciously: Distinction from lower perceived classes (cf. Weber 1980, 179). This form of discrimination has developed over long periods of time (cf. Halfdanarson/ Vilhelmssson 2016, 33f). This leads to an internalisation and recognition of structures of denial with the feelings of devaluation (Kemper/Weinbach 2016, 51).

Due to the clear demarcation, people from lower income strata acquire negative self-concepts more frequently and the development is reinforced via advertising (cf. Dörre/ Scherschel/ Booth/ Haubner/ Marquardsen/ Schierhorn 2013, 198). The feeling of being left behind, despite the equal living conditions, triggers problem spirals that make the situation more precarious, both professionally and in terms of partnership (cf. ib.; cf. Dackweiler/ Rau/ Schäfer 2020, 111). Real status affinity increases that poorer people feel devalued because they are suggested by the upper class to be without inferiority. The assessment is shared that "the consequences of this inferiority souffléed into workers and workers' children for decades and centuries" could turn into aggression (Glawion 2019, 62; cf. Häberlin 1947, 60).

For this reason, it should be shown what a politically explosive power classism possesses: discrimination against lower classes (downward classism) generates post-haste upward classism, which based on jealousy and envy (Liu 2011, 199f). Thus, prejudices against the rich and intellectuals also become entrenched from affected people of the working class (Kemper/ Weinbach 2016, 23). The social division reinforces each other and also solidifies intersectional sexism and racism (cf. Zitelmann 2019, 38). However, downward and upward classism can't be equated, since "only the dominant groups have the power to make their prejudices structurally effective by means of oppression" (Kemper/ Weinbach 2016, 51).

2.2.   Connecting points in the mobility sector

In the field of mobility, there are several representations of how classism might present itself. First of all, the most visible form of classism should be pointed out. Classism is consciously perceived when it is exposed (cf. Baron 2014, 226). This includes, for example, "setting cars on fire", "destroying prestigious luxury objects", "physically blocking access" or "occupying status objects". These are illegal activities ranging from civil disobedience to the use of violence against objects. Media attention is particularly high when physical violence is used against objects and people, regardless of the nature of the problem. It is uncertain whether the actions can be attributed to the spectrum of left-wing extremism or whether they are carried out independently of political convictions, i.e. are "structurally" conditioned (cf. Treskow/ Baier 2020, 22; cf. Riekenberg 2008, 175). According to Treskow and Baier (2020), radical left-wing activists are enrolled at universities and thus tend to belong to the highly educated classes; the existing potential for violence of discriminated and frustrated people due to precarious living conditions is not based on political but socially structured motives (cf. Treskow/ Baier 2020, 42).

In the case of downward classism, which is more invisible, the mind-set matters. It happens that disadvantaged people find out about the superiority thinking strategy of their counterparts, which devalues them. In most cases, it doesn't happen directly, but indirectly, avoiding contact. Driving up in an expensive car, overusing space with the vehicle, using luxury vehicles such as private jets and yachts has one goal: isolation from society and avoidance of poorer people, which fuels prejudice against them, which is reproduced and maintained in their bubble (cf. Allports 1954). But is it the free choice of each individual? Sociology has shown that freedom is unequally distributed. It is closely linked to social and monetarist status, which is already unequally distributed to begin with. Hanging out alienated prestige objects such as a luxurious car is therefore a classist action that makes visible the lack of freedom of others and wants to provoke them through unjust conditions. Accordingly, it's no surprise that those who aspire to a higher level of narcissistic admiration are described from the outside as arrogant, wasteful, stingy and dishonest (cf. Leckert/ Richter/ Schröder/ Küfner/ Grabka/ Back 2018, 773ff). It is the result of seven experiments by US researchers that the upper class more often spreads untruths, cheats more often and takes the right of way from others in road traffic (cf. Piff/ Stancato/ Côté/ Mendoza-Denton/ Keltner 2012, 4087f). The psychologists assume that members of higher social classes probably behave more immorally because greed is seen in a more positive light in this part of society (cf. Pfiff et al. 2012, 4089). The psychological studies offer clues as to how it can be possible to uncover classism.

3. Research methods

In the research method, it is necessary to choose how classism can be detected. In many studies on racism, the psychological component plays a crucial role. An attempt will be made to reveal attitudes and classist thought patterns. Because only the dominant groups have the power to make their prejudices structurally effective by means of oppression, this study only handles with downward classism.

3.1. Social psychological questionnaire based on the GMF model

Social psychological studies can measure explicit prejudice. The explicit measures of social distance are used in these studies (Correll 2010, 47f). From social psychology, a standardised survey of hundred people could be created with question programmes. Facts of classism can be categorised individually. "Negative prejudice", "ability to indulge", items about status and status symbol, "inacceptance of equal living conditions" and finally "encapsulation" are the categories. The respondents indicate how they feel about different facts (cf. Correll 2010, 47). This is about the factual situations in which the respondents show tolerance and understanding or behave in an intolerant, resistant and uncomfortable way. The study focuses solely on the emotional component (feelings of sympathy or antipathy), so inner psychological categories like narcissism or inferiority complexes can be regarded.. There are items that reveal general classism, others measure personal prestige affinity to vehicles. The exact group of people in question should be high status or rich. Specifically, however, stigmatising prejudice statements must be embedded. The influential GMF model of Wilhelm Heitmeyer, a right-wing extremism researcher, is used internationally in studies on group-related misanthropy (cf. Zick/ Küpper/ Hövermann 2011). The GMF model conceptualises group-based misanthropy as consisting of several individual dimensions, all of which are strongly interconnected (cf. Nolden/ Sudnik 2020, 13). The dimension of "devaluation through classism" of the GMF syndrome must be supplemented by mobility-independent items. The selection of the group of people is oriented towards drivers of expensive car brands, users of private jets or yacht owners. A basic assumption is that the use of luxury vehicles also means financial wealth. Contact with the group of people can be established through salespeople. The survey can be conducted online.

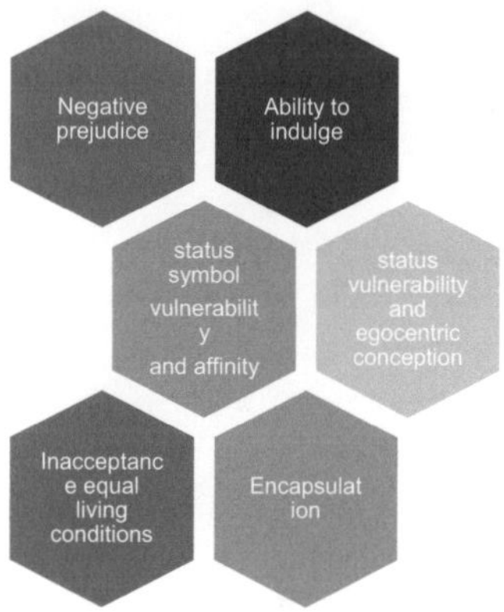

Figure 1: Categories of classism in the social psychological questionnaire

3.2. Qualitative semi-standardised interview

In a large European anti-discrimination study, there is a need to record data on discrimination by authorities (cf. Ahyoud 2018, 28/ cf. Zick/ Küpper/ Hövermann 2011). Even it's criticised that such an indicator is hardly representative of the discrimination and that classism is not even mentioned in this study, a new database could be created and bring a solution. The scientific goal of making classism measurable can be realised through a proactive interview study and capture "complaints from victims". It isn't enough to set up a passive anti-discrimination office for classism, which waits for reports, as this form is still one of the least researched types of discrimination in the European area. Disclosure can work through a proactive interview study, which should actively seek out affected people. Where should people be selected and interviewed? A sole selection of social hotspots or stigmatising poor-looking people would not be sufficient for the study, because victims of classism can be suspected in every stratum and, on the other hand, the study doesn't conduct any foreign categorisation via status profiling for ethical reasons (cf. Simon 2012, 1386). Travelling by train in Germany isn't a privilege but largely a means of transport used by society as a whole, so train riders meet the requirement of a heterogeneous longitudinal section of the population. Interviewers who are able to address hundred people on the spot and create an appropriate atmosphere will be necessary for the implementation. In contrast to the first method, in which people are obviously looking for their own discriminatory ways of thinking, the interview study is about experienced discrimination. Furthermore, a multidimensionality is shown by including the mobility situation, namely in the question whether travelling by train is considered inferior and also whether classism is institutionalised, for example in ticket controls. The interview should be only semi-standardised, in which guiding questions are mentioned, but also open to personal experiences with classism.

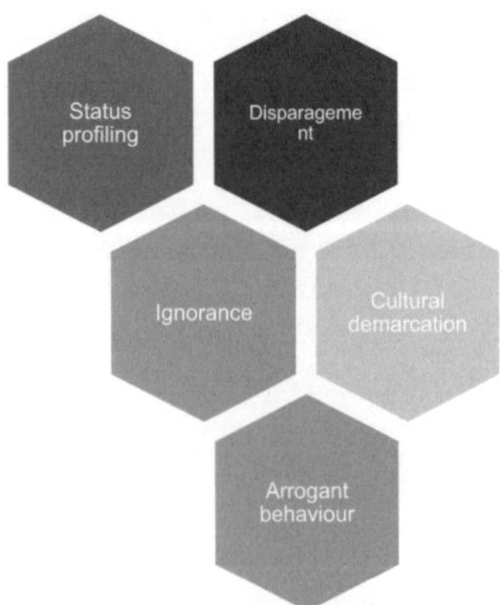

Figure 2: Categories of classism in the semi-standardized interview

4. Discussion about evaluation

Finally it should be discussed if the methods are able to detect classism. Although classism is still a neglected field in the field of sociology, there is a fundamental potential to reveal classism with both methods and to show it as a form of discrimination. With the first study, it could be shown that classist thought patterns are obviously lived out and can be linked to many inner-psychological problems. It can be assumed that it will not only show status thinking in the form of certain means of transport, but also the devaluing and discrimination against poorer people. However, Bogner and Landrock's (2015) fear that the test persons will only give socially accepted answers must be taken into account. It will be a challenge to get the study to people under the banner of anti-discrimination research; possibly one or the other salesperson will have sales-strategic concerns. However, a covert form of research, e.g. embedding some car design or driving comfort related items, is not considered necessary as classism is still too unknown.

The second method, which involves greater effort, can be qualitatively more detailed. It can be assumed that classism is able to be addressed much more frequently if one can speak from a participant role. An attempt has been made to categorise the forms of discrimination: The one against rail passengers, the other one as a localised form, institutionalised discrimination and those with cross-location factors. So it is possible to show the explosive nature of classism as a phenomenon in society as a whole, and on the other hand to establish the relationship of classism to a devaluation of the train passenger, which can contribute to an understanding of the unequal treatment of different vehicles. The semi-standardised interview doesn't use a scoring system, but classism can be detected in a non-hierarchical way at various points.

It is expected that high status affinity over expensive vehicles is closely related to downward classism and minority complexes. In the second study, the breadth of the spectrum of classism increases. If Kemper and Weinbach (2016) are right, who speak of an internalisation of classism, high scores can be expected, maybe even record scores in anti-discrimination research. With such an expectation, this empirically founded study would have been able to demonstrate that classism produces unequal living conditions and status-related vehicles affinity – a possible appeal to confront classism consistently.

List of figures

Literacy

All links were checked for functionality on 18 December 2021.

Allports, Gordon (1954). The Nature of Prejudice. Addison-Wesley. Reading.

Ayound, Nasiha (2018). Welche Arten von Antidiskriminierungs- und Gleichstellungsdaten gibt es? In: Ahyoud, Nasiha/ Aikins, Joshua Kwesi/ Bartsch, Samera/ Bechert, Naomi/ Gyamerah, Daniel/ Wagner, Lucienne (2018). Wer nicht gezählt wird, zählt nicht. Antidiskriminierungs- und Gleichstellungsdaten in der Einwanderungsgesellschaft – eine anwendungsorientierte Einführung. Vielfalt entscheidet. Diversity in Leadership. Citizens For Europe. Berlin. pp. 25-28.

Baron, Christian (2014). Klasse und Klassismus - Eine kritische Bestandsaufnahme. In: PROKLA. Verlag Westfälisches Dampfboot. Vol. 44. Issue 175. Nr. 2. pp. 225-235.

Bogner, Kathrin/ Landrock, Uta (2015). GESIS Survey Guidelines - Antworttendenzen in standardisierten Umfragen. Leibniz Institute for Social Science. Mannheim.

Correll, Joshua (2010). Measuring Prejudice, Stereotypes and Discrimination. In: The SAGE Handbook of Prejudice. Stereotyping and Discrimination. Vol. 1. Los Angeles. pp. 45-62.

Dackweiler, Regina-Maria/ Rau, Alexandra/ Schäfer, Reinhild (2020). Frauen und Armut - Feministische Perspektiven. Verlag Barbara Budrich. Berlin/ Toronto.

Dörre, Klaus/ Scherschel, Karin/ Booth, Melanie/ Haubner, Tine/ Marquardsen, Kai/ Schierhorn, Karen (2013). Bewährungsproben für die Unterschicht? Soziale Folgen aktivierender Arbeitsmarktpolitik. Campus-Verlag. Francfort a. Main/ New York.

Glawion, Sven (2019). Als Arbeitertochter unter Marx' Erben: Eine Lesart zu „Klassenliebe" von Karin Struck. In: Revista de Estudos Alemaes. Vol. 8. Is. 5. Universidade de Brasilia Nücleo de Estudos de Linguas e Culturas Germânicas. Brasilia. pp. 45- 72.

Häberlin, Paul (1947). Minderwertigkeitsgefühle. Wesen, Entstehung, Verhütung, Überwindung. Schweizer Spiegel Verlag. Zurich.

Halfdanarson, Guðmundur/ Vilhelmssson, Vilhelm (2016). Historische Diskriminierungsforschung. In: Scherr, Albert (Hrsg) (2016). Handbuch Diskriminierung. Springer. Wiesbaden. pp. 25-37.

Jensen, Barbara (2012). Reading Classes: On Culture and Classism in America. Cornell University Press. Cornell.

Kemper, Andreas/ Weinbach, Heike (2016). Klassismus. Eine Einführung. 2nd Edition. Unrast-Verlag. Munster.

Knierim, Bernhard (2016). Ohne Auto leben. Promedia. Vienna.

Leckert, Marius/ Richter, David/ Schröder, Carsten/ Küfner, Albrecht/ Grabka, Markus/ Back, Mitja (2018). The rich are different: Unravelling the perceived and self-reported personality profiles of high-net-worth individuals. In: Psychology. Vol. 110. Issue 4. pp. 769-789.

Leondar-Wright, Betsy (2005). Class Matters: Cross-Class Alliance Building for Middle Class Activists. New Society Publishers. Gabriola Island.

Liu, William (2011). Social class and classism in the helping professions: Research, theory, and practice. Sage Press. London/Singapore/New Delhi/ Thousand Oaks.

Nolden, Dorothea/ Supik, Linda (2020). Analyse der Forschungsbefunde zu antiziganistischen Einstellungen in der deutschen Bevölkerung Forschungsstand, Kritik, Alternativen. URL: https://www.institut-fuer-menschenrechte.de/fileadmin/Redaktion/PDF/UKA/Analyse_der_Forschungsbefunde_zu_antiziganistischen_Einstellungen_in_der_deutschen_Bevoelkerung.pdf

Piff, Paul/ Stancato, Daniel/ Côté, Stéphane/ Mendoza-Denton, Rudolfo/ Keltner, Dacher (2012). Higher social class predicts increased unethical behaviour. University of Michigan. In: Proceedings of the National Academy of Sciences. Vol. 109. Issue 11. pp. 4086-4091.

Riekenberg, Michael (2008). Auf dem Holzweg? Über Johan Galtungs Begriff der „strukturellen Gewalt". In: Zeithistorische Forschungen/Studies in Contemporary. History 5. pp. 172-177.

Scheytt, Carla (2020). Forschungsethik in der qualitativen Sozialforschung. In: RUB Methodenzentrum. University of Bochum. URL: https://methodenzentrum.ruhr-uni-bochum.de/e-learning/forschungsethik-in-der-qualitativen-sozialforschung/

Simon, Patrick (2012). Collecting ethnic statistics in Europe: A review. In: Ethnic and Racial Studies. Vol. 35. Issue 8. pp. 1366-1391.

Treskow, Laura/ Baier, Dirk (2020). Wissenschaftliche Analyse zum Phänomen des Linksextremismus in Niedersachsen, seiner sozialwissenschaftlichen Erfassung sowie seiner generellen und spezifischen Prävention. Criminological Research Institute of Lower Saxony. Institute for Delinquency and Crime Prevention Department of Social Work. Zurich University of Applied Sciences. Hanover/ Zurich.

Weber, Max (1980). Wirtschaft und Gesellschaft. 5th Edition. Mohr. Tuebingen.

Wondratschek, Florian (2020). Durch Klassendenken zur großen Liebe? Interdisziplinäre Analyse über Klassismus in der Partnerwahl. Institute for special need education. University of Education Ludwigsburg. In: Wondratschek, Florian (2020). Klassismus in der Partnerwahl. Durch Klassendenken zur großen Liebe. Grin. Munich.

Zick, Andreas/ Küpper, Beate/ Hövermann, Andreas (2011). Die Abwertung der Anderen. Eine europäische Zustandsbeschreibung zu Intoleranz, Vorurteilen und Diskriminierung. Friedrich-Ebert-Stiftung. Berlin.

Zitelmann, Rainer (2019). Die Gesellschaft und ihre Reichen: Vorurteile über eine beneidete Minderheit. FinanzBuch Verlag. Munich.

Appendix

I.   Social psychological questionnaire based on the GMF model

On classism in mobility, the following questions could be imagined for car drivers:

(1) "Would you like it, wouldn't you care, or wouldn't you like it so much if there were several sports cars parked next to your home?"

(2) "Would you like it, would you not care, or would you not like it so much if your neighbour drove a sports car?"

(3) "Using a scale of 1 to 10, please say how comfortable you would feel about a poorer man taking pictures of your car.?"

(4) "Using a scale of 1 to 10, please say how sour you would feel about people criticizing your car."

(5) "Using a scale of 1 to 10, please say how proud you would feel to drive a sports car."

(6) "On a scale of 1 to 10, please say how annoyed you are with people saying you have a privileged lifestyle"

(7) "Often financially poor people say they would like to have more money to travel. Using a scale of 1 to 10, please say how self-inflicted you think the poverty of such people is"

(8) "Using a scale of 1 to 10, please say how unhappy you would become if a poorer rated neighbour suddenly flaunted a luxurious lifestyle."

(9) On a scale of 1 to 10, please say how justified you think it is to say that doctors, lawyers and managers have earned a higher lifestyle, e.g. through more air travel, more expensive cars and real estate compared to other groups?

Downward classism - Clear stigmatisation issues

(C1) "Would you have problems travelling to work in a vehicle with the lower class?"

(C2) "Do you think poor people should rather keep to themselves?"

(C3) "Do you think poorer people tend to commit more crime?"

N.	Main topic	Point allocation	Points
(1)	Ability to indulge	Like it = 5 Wouldn't care= 0 Don't like = 5	
(2)	Ability to indulge	Like it = 5 Wouldn't care= 0 Don't like = 5	
(3)	Negative prejudice	10=1, 9=2,…	
(4)	Vulnerability of the status symbol	1=1, 2=2,…	
(5)	Status symbol affinity	1=1, 2=2,…	
(6)	Vulnerability of the status	1=1, 2=2,…	
(7)	Egocentric conception about status	1=1, 2=2,…	
(8)	Inacceptance equal living conditions	1=1, 2=2,…	
(9)	Inacceptance equal living conditions	1=1, 2=2,…	
(C1)	Encapsulation	Yes = 35 No = 0	
(C2)	Encapsulation	Yes = 35 No = 0	
(C3)	Negative prejudice	Yes = 35 No = 0	

35 points or more, the person has discriminatory attitudes in the form of classism.

Below the value, further tendencies can be recognised:

Inferiority complexes could be disclosed in the case of a high score on (1), (2), (4), (5), (6), (8) and (9). A form of narcissism could be disclosed of a high score on (3), (6), (7) and (9). If the ability of indulge would be low, but the inacceptance of equal living conditions, the person would have also discriminatory attitudes. Only a high score in (5) wouldn't express a form of classism, but an enthusiasm for sports cars.

The interview should be only semi-standardised, with guiding questions, but also open to personal experiences of classism.

(1) "Do you feel or know any relatives who are ignored because of having less money?"

(2) "Do you think people here in [location name] feel similarly about this?" (reasons?)

(3) "When travelling by train, have you experienced someone being belittled and can you describe it?"

(4) "Have you ever had the feeling of being labelled poor by other travellers?"

(5) "Did anyone ever make you feel inferior because you travel by train?"

(6) "Did you ever find yourself unable to take part in conversations because the person you were talking about things that only affect rich people?"

(7) "Have you ever experienced a rail employee mocking fare dodgers?"

(8) "Do you experience debates on your train journey in which third parties are blasphemed about the financial situation?"

(9) "Between you and me, do you experience such debates in private?"

(10) "Are there people who give special preference to people in their immediate circle because they are rich?"

(11) "Do you remember a particular commuting situation during our trip where you were upset by a person's arrogant behaviour?"

N.	Main topic	Classism in the mobility sector	Rating
(1)	Ignorance		Possible discrimination on the basis of capital
(2)	Ignorance		Possible prejudices/ experiences of classism in that location
(3)	Disparagement	x	Possible discrimination on the basis of status
(4)	Status profiling	x	Experienced discrimination on status profiling

N.	Main topic	Classism in the mobility sector	Rating
(5)	Disparagement	x	Rail travel is devalued, possible discrimination on the basis of status
(6)	Cultural demarcation		Encapsulation, dependent on intention
(7)	Disparagement	x	Possible experience of institutionalised classism
(8)	Disparagement	x	Possible discrimination on the basis of capital
(9)	Disparagement		Possible discrimination on the basis of capital
(10)	Status profiling		Possible positive discrimination on the basis of status
(11)	Arrogant behaviour	x	Possible experiences of arrogance based on classism

In qualitative studies, it is relevant to examine the cause in detail. If money and status are the trigger point of discrimination, it can be mentioned classism. The more often those motives play a role, the more extreme classism is manifested. One answer can be enough, that a person was discriminated in the form of classism. For analysing the mobility sector externally, a confirmed answer to (5) would show also a inferior-thinking of a sustainable collectivistic vehicle in connection to a person. A classism that may has created many inequalities in transport. (7) checks, if there might be institutional classism in the current train system. The other items about the mobility sector are limited only locally in the rail travel.

YOUR KNOWLEDGE HAS VALUE

- We will publish your bachelor's and
  master's thesis, essays and papers

- Your own eBook and book -
  sold worldwide in all relevant shops

- Earn money with each sale

Upload your text at www.GRIN.com
and publish for free